The Little Handbook of
Symbiotic Agriculture

Colette Mourey

1

ISBN: 9781973294283

The Little Handbook of
Symbiotic Agriculture
— by —
Colette Mourey

Translated by
Astra d'Oudney
www.scorpiotraduction.com

— published by —
A.D.C.M. on KDP

https://fr.wikipedia.org/wiki/Colette_Mourey

Website :

http://colette-mourey.com/

AMERICAN ENGLISH SPELLING

Dedication

For Philippe

TABLE OF CONTENTS

Preface

This Little Handbook provides a straightforward definition of the philosophy of agronomic *symbiotics*, as directly inferred from respectful observation of *natural, spontaneous relationships*.

Within this context, we touch on *"soil-fauna-flora"* systems which are so closely intertwined that they prove absolutely inseparable. One immediately notices that they generate complex *languages*, and that, as they are possessed of *memory*, they are capable of *enculturation*. The infinite richness of this creates veritable *civilizations*.

We owe a threefold, sincere deference to the intelligent soil (which is the *sine qua non* of all life on Earth), as well as to plants and animals fully immersed in the flow of life, and so intricately intuitive. All the kingdoms of nature turn out to be of one identical essence and of equal worth to that which we bestow on our own bodies and lives.

It is in this regard that cultivating, growing and raising… all come down to *love*. That is to say, recognizing and serving that vital impetus which carries us along and of which we are, by no means, the only fine expression.

All realms demonstrate a complex social life. All accumulate valuable knowledge. All are steeped in a greater collective universal memory which they help to nurture and enrich.

We owe it to ourselves to work *with our Earth* and not constantly *against it.*

This will be made possible by close observation of continual interactions, and sharing and cooperation mechanisms which we must henceforth refrain from throwing off balance.

Let us create this vital symbiosis between humankind and all of the kingdoms from which it originates and feeds off, in acknowledgement of the value and essential nature of every one of them.

Let us love the organism formed by planet Earth, and which is itself incorporated into an ever vaster interlacing of organisms – a unitary lattice from the microscopic to the macroscopic level.

Chapter One
By Way of Introduction
What Is "Symbiotic Agriculture"?

Symbiotic Agriculture is almost diametrically opposed to what we anthropomorphically and aggressively call "agriculture" today. It involves rediscovering, fostering and respecting the ***primordial relationships between plants, animals and minerals***. Unceasingly, over the eon-long course of Evolution, these have come about and developed mysteriously and ***spontaneously***.

It turns out that ***all biotopes result harmoniously and harmonically from a unitary lattice of interlinked data*** whose common objective is ***sharing and cooperation.***

Possessing highly efficient communication, these ***autonomous, hierarchically-structured microsocieties,*** within which all realms of living things intertwine, are organized ***instinctively***. Associations and dissociations take place through complex biochemical

7

dialogue, with an attractive or repulsive function.

All of these interlocking organisms complement one another freely. They share their defenses; assist each other on a constant basis; and mutually enrich themselves, whilst diversifying immeasurably, as a result, the *genetic heritage* of the biotope they inhabit.

These are veritable *micro-civilizations* with languages, demonstrating *culture and memory,* and exhibiting profound shared intelligence. They interconnect beings which are fundamentally *indissociable* from one another, without requiring any involvement from us.

So, we no longer need to continue damaging the earth – that incredibly fragile humus – with inappropriate plowing combined with a proliferation of harmful treatments. The pointlessness of this is now well established, as we shall detail in this book.

Everything is useful – necessary and sufficient – in the architecture of landscapes, as in *natural, instinctive groupings.* There are no such things as "weeds" or unnecessary predators. Any environment naturally tends towards balance by using the synergy derived from its own strengths.

Symbiotic agriculture is based on keen, meticulous and *scrupulous observation* of *natural, untouched landscapes.* We have *everything* to learn from them. How, in the absence of Man, does our planet live, resonate, channel vital energy, attune itself and adapt in complete self-sufficiency and autonomy?

First of all, let us remember that every ecosystem sets itself up like an *algorithm*; a unitary assembly of data whose gradual realization demonstrates the archetypal omnivalence of an unchangeable invariant. The latter, however, turns out to be basically mobile, open and dynamic through its continual, simultaneous inclusion of wider, more general information.

Every environment originates and acquires its longevity from a founding pre-encoding of ideas and languages which are

9

instantly translated into dynamic messages. These make it possible for various architectures to restructure themselves collectively by recombination, evolving through each other, and refining their knowledge in perfect symbiosis.

Although the aforesaid interrelation will certainly develop between plants instinctively, it occurs mostly as part of the unique interactions which take hold within the whole "soil-plant-animal" system. They especially involve bacteria, fungi and algae. These continual reciprocal contributions are underpinned by the immense biodiversity of a plentiful soil fauna – billions of micro-organisms which inhabit the top layer of our arable land. They feed and aerate that indispensable stratum from which we draw our vital force and which we harm all too frequently...

When vegetation comes out autonomously in an environment where ideal spontaneous symbiosis is respected, the soil is amazingly full of life!

Consequently, as proof of this, we can present the analysis carried out on its fine composition. In symbiotic agriculture, we immediately notice the **extremely high density of organisms which populate**, oxygenate and soften the soil, at the same time as feeding from and nourishing it continually with their dead bodies and excreta. In return, these provide plant cover with invaluable nitrogen and phosphorous complexes. The ceaseless activity of these organisms creates the best naturally compatible "manure" for the primordial biotope of the location in question.

So, everything naturally returns to the earth which will use it to establish a yet better balance, all the while drawing on it to nurture an abundance of life itself!

To begin with, we may consider that any landscape we analyze (substratum and environment in the broadest sense of the term) is an exact translation of an algorithm. It is the concrete expression by a group of creatures at a given moment of the sole underlying *information network* and completely seamless architecture. Let us recall the universality of series, such as the Fibonacci sequence, which is obeyed by all plant growth.

It appears that although every environment is fundamentally *unique* and *autonomous* at the same time, it is *not autarkic*. Different biotopes *entangle their own algorithms* and harmonize at sequential intervals, helping each other, sharing and cooperating unceasingly.

On the one hand, the earth is harmed by plowing too deep, with it becoming compacted, broken down, dried out and inevitably dying under the combined influence of frost and sun, from which it has lost all natural protection. On the other hand, the humus is soft and full of life, vibrant, adaptable, naturally irrigated (water management is organized in the most efficient and responsible way possible), thus indicating a *complex intelligent civilization wherein mineral and organic realms intertwine,* and which is instinctively sustainable by itself.

One only has to study the soil and configuration of our forests to confirm this. Their primordial ingenuity makes it possible for plants, micro-organisms and macro-organizations of all types skillfully combined to embrace one another harmoniously (and harmonically) from the root system up to the crown in a single unitary breath!

It is for this reason that, before we even think of sowing and planting, we absolutely must *analyze the reflexes of the plant in the wild*; learn *what its natural companions are*; become acquainted with *its instinctive symbioses*; and exhaustively pinpoint its role and *precise ecological function*. The aforegoing ensures that our actions move exclusively in the same direction as the forces of life have naturally headed from the beginning. This is the only way to achieve a crop with an attractive yield, because organisms' propensities will have been respected and they will *thrive healthily* as a result.

A plant – as with any entity which life has created – sets itself up in its own sphere like a *specific sum of information*, stemming, for the most part, from unitary genetic pre-encoding. This derives directly from the vital energy which will have stimulated its growth. Hence, for example, its etheric form will have appeared well before the physical manifestation, which is subordinate to it.

Moreover, the aforementioned specific system can only be constructed if it is in perfect *interrelation* with more extensive structures. This will expand foundations, whilst bringing about, in response, intense *communication* activity, capable of feeding all *telluric* and, more broadly, universal *memory*.

This is an essential condition for the good health of any organism, whatever realm it belongs to! *A bodily envelope is both a unitary structure and the smallest unit of a vast, complex social organization.* The same applies to humans, animals, minerals and plants!

The phenomenon is identical for all micro-symbiosis, through which single-celled beings, combining all realms, are brought together and organized into a hierarchy. They all have *memory* and are carriers of the same vital clairvoyance. At the same time, micro-symbiosis is directly operational in concrete terms and unlimited in its ideational implications as much as in its long-term impacts.

An isolated plant separated by a furrow's length from the only counterparts it has, looks like a prisoner in a concentration (or extermination!) camp. Meanwhile, it immediately loses *the coherence of its primordial information,* which comprises its irreplaceable identity. Similarly, because this layout prevents the plant from interacting with its natural, fellow organisms, it becomes literally autistic. It would be quite an understatement to say that the plant is suffering!

From this, we may deduce that when we create uniform areas of intensive monoculture, we directly abuse the forces of life and irremediably attack Earth. What is more, we are reaching the point of no return – this has become evident on a global scale!

15

At the same time, we cannot but stress the *current unfathomable despair of the impoverished, agricultural community who are exploited and unjustly treated!*

If we continue applying our industrial methods to farming, all signals will turn red. Aside from the immeasurable damage of uncontrolled pollution, we are on the eve of the greatest world famine we will have ever known and had to contend with since ancient times! We will have caused it from start to finish!

Thus, it must be concluded that the *agriculture of tomorrow has to be developed to be the exact opposite of what we call "agriculture" today.*

At present, because we cultivate vast, single areas which we have flattened, cleared and made uniform, we are forced to spread heaps of fertilizer and treatments on them, literally *poisoning* the substratum upwards to the air above. All we can possibly get are vegetables which are essentially as sick and unwholesome as the animals we raise. These

16

too are bred in unspeakable concentration-camp like systems.

Let us consider that, for our part, each one of us is a piece of specific unitary information, inextricably interwoven with other entities who are similarly built structurally, all originating from generally accepted and respected languages and cultures, within harmoniously balanced surroundings. Therefore, we must remember that what appears to be a distinctive feature of Man is also the vital configuration of healthy minerals, plants and animals.

Symbiotic agriculture makes it possible to sustain spaces in which we find the greatest biodiversity, as natural abundance – mineral, plant and animal – is considered to be an indivisible whole which is inherently sharing and cooperative, without having any external orientation placed upon it. *Every species therein naturally protects and strengthens its neighbors.*

That is why, in the long term, we will not fail to see the *close resemblance which will develop between the "natural" landscape*

17

(of a primordial, instinctive and autonomous structure) and the agricultural landscape which Man will gradually put together innovatively, in full observance of the dynamics of vital forces (such momentum developing logically along the lines of the former).

As for **optimum symbioses** between various realms, these are easy to pinpoint and encourage. They are to be seen in primordial landscapes. We shall gradually find them there through simple, attentive and continual observation.

What About Crops?

Actually, we might smile at this thought. When we need to go and get our supplies from such rural areas, we might feel a bit flummoxed and flee from the task!

However, are we not in the digital, IT age? It would be so simple using existing machines, equipped with suitable software, to pick or pluck produce in a highly *selective* manner! In fact, to economize energy, a crop of two, three, or more species could be sorted straightaway and be put into a caddy with compartments.

What about the farming "profession"? If all we do is reproduce nature, won't farmers disappear?

On the contrary, they will become more specialized than ever. *They will be able to carry out exhaustive analyses; work whilst respecting vital energy; learn to gauge carefully how much soil enrichment product to use on primordial structures; remain constantly reflective and pursue research.* In this context, all the aforegoing makes the job of farmer of the greatest use; the finest and most complex of professions!

It will be a job requiring immense knowledge. This will be acquired through ongoing training and rely on an advanced *research "hub"* in constant intercommunication between science and farming techniques.

That is why the low incomes which undermine the work and health of our farmers will no longer be tolerated. They will automatically receive executive-class wages in keeping with far greater skills and responsibilities.

In these troubled times, is not the farmer the guardian of our Earth; the last bastion who will be able to stand up against all forms of imbalance, including climate change? He or she alone will be the one to ensure the sustainability of the human race, as well as all life on Earth. This deeply affects future generations. *It is our own children who will be saved by symbiotic agriculture from the current scheduled disaster!*

Chapter Two
Holism and Symbiosis

Symbiosis always develops within globality; a structured *totality,* in full, complete wholeness. This occurs through an *invariant* which allows the combination of immutability and unique dynamism, arising from a free openness to all similarly constituted totalities.

Symbiosis is the hallmark of this fundamental life-defining *holism,* although it cannot be reduced down to a few fragmented and accordingly inaccurate images which we like to contrive for ourselves, either out of expediency or greed.

It is expressed **simultaneously from the macroscopic down to the microscopic levels,** arising from primordial, complex and evolving communication systems.

On the one hand, we, as *unitarily informed organisms,* are not able to feed ourselves exclusively with *exhaustively informed organisms,* if we want to sustain optimum health for ourselves.

On the other hand, an organism – of any kind – can only truly be "informed" when it lives in *symbiosis* with the abundantly whole environment which suits it. It is from that environment that it will elicit all of the ramifications of its identity, as with all of its unique characteristics.

Plants are crippled within a biotope which has been depleted by Man's draconian choices. They can no longer access the data which are essential for them to thrive, and accordingly, they will no longer have the nutritional properties they might have had.

We eat ersatz fruit and vegetables all too often!

Therefore, we need to accommodate a *primordial landscape as a whole*; not glimpsed as a sketchy outline.

Symbiosis uses all means at its disposal through unitary essence, simultaneously harnessing *all kingdoms* and their instinctive interrelations equally.

The parameters of the aforesaid symbiosis must absolutely be holistic – inextricably *mineral, animal and plant.*

a) Mineral Parameters

Loose soil is a delicate surface structure resulting from bedrock breaking down into mineral particles over the course of millennia. It is a complex, dynamic environment, which either naturally fosters germination (as soon as there is a correlation between the genetic heritage of the future plant and outside conditions) or forces the seeds to remain *dormant*.

Whilst it is defined by its long history, it is also constantly rebalancing itself.

Analysis can determine its specific texture (varying degrees of coarse or fine sand, alluvium, clay); the structure (gritty, lamellar, solid, etc.); density (varying degrees of compactness or porosity, etc.); natural water-draining capacity; pH level (acid, neutral, base), and so on.

This fragile substratum is the irreplaceable haven for microbial, animal and plant biodiversity. It plays the role of a *dynamic interface between the lithosphere and biosphere* and is the result of continual, complex interchange. At any specific moment, it turns out that its nature, composition and pH are entirely dependent on the biodiversity of a mini-society of micro-organisms, interconnected with the flora and fauna which will have settled in the biotope thus created. This will be reflected in the bioindicator species which will flourish there naturally.

We fully concur with the adage, "healthy soil for a healthy life"!

1) Nature

Nature's coarse materials (sand, alluvium and clay) create an intelligent water distribution system, while fine particles (mineral colloids) primarily are chemically and biologically active.

Moving on, the soil's organic fraction, called humus, is a *structure of carbonaceous material* derived from both the decomposition of plant litter (root exudates) and the activity and metabolism of chiefly fungal, algal and bacterial organisms.

24

In all cases, the various types of soil have an autonomously instinctive tendency to balance each other, through synergies combining all of the resources at their disposal. This applies especially to conglomerates resulting from mineral, animal and plant inputs enriching them and which they help nurture.

With just a quick glance, we can identify a substratum's nature by careful observation of the whole body of bioindicators.

1) Sandy Soil

Sandy soil is made up of relatively sparse and varying sizes of coarse-grained silica. It has a very loose and unstable structure making it extremely prone to wind erosion, for example. Also, it is basically pervious to water and organic matter. Bioindicator plants for this type of soil include, mugwort, couch grass, common mallow, black nightshade, common yellow woodsorrel, common chickenweed, and so on.

Retaining what used to be called "weeds" makes it possible for the soil to enrich itself naturally, without any risk of one species becoming overdominant. This is because growth takes place *as a whole, in constant interrelation,* with proportions changing as the substratum balances itself.

The aforementioned bioindicator plants will immediately have a positive influence on plants which will be added.

Additionally, we must stress the absolutely indispensable role of natural forest dwellers. This is demonstrated by the unique benefits proffered to all types of substrata and crops by the continued presence of hedges, copses and woods.

Many trees flourish on sandy soil, fertilizing it in return over the course of centuries. For example, groups comprised of acacias, alder, birch, sweet chestnut, oak, maple, ash, horse chestnut, walnut... larch, pine, sequoia, cypress... Thriving between these trees, shrubs complement their action: heather, broom, hazel, and so on.

2) Silty Soil

Silty soil, rich in alluvium, is unstable and fragile, although it is more highly water-retentive than sandy soil. This biotope's natural flora can include, mugwort, buddleia, poppies, thistle, convolvulus, and so on.

3) Clay Soil

Clay soils often get heavy, sticky and so saturated that they become impervious, making them the last kind of soil to warm up in spring. Amongst other plants naturally occurring on this type of soil, we find orache, wild garlic, thistle, convolvulus, plantain, buttercup, dandelion, coltsfoot, marigold, fumaria, wild tulip, Grey-Field speedwell, and so on.

2) pH

However, *the most important parameter for soil is its pH*, resulting from the aforesaid type. Neutral soil has a pH of 7. Measurements below this indicate acid soil and above, alkaline soil. Here again, a soil's pH leads to a specific bioindicator flora.

For instance, on acid soil, we find heather, ferns, common broom, fox-and-cubs, common yellow woodsorrel, sheep's sorrel, dandelion, gorse, blueberry, sweet chestnut, and so on.

A neutral environment will be host to sedge, wood speedwell, enchanter's nightshade, stitchwort, and so on.

At the other end of the scale in chalky soil, we have an abundance of bellflowers, knapweeds, clematis, common chicory, meadow sage, hellebore, field mustard, salad burnet, elderberry, elm, dogwood, field maple, cherry, and so on.

Bioindicator plants are a faithful witness to all a substratum's workings.

For example, poorly-drained soil leading to an excess of water will result in such self-propagating plants as, field horsetail, creeping buttercup, coltsfoot, golden rod, etc.

Thus, a total symbiosis between the flora and soil, which turns out to be infinitely more alive and responsive than we had previously thought, is constantly observed.

This is what we call the *"soil-plant"* system, which reflects *a complex language, underlying memory and a boundless ability for enculturation...*

b) Organic Parameters

Plants benefit from the rhizosphere which comprises an environment of extremely rich organic diversity wherein micro-fauna and micro-flora harmoniously co-exist (eukaryotes: fungi, algae, protozoons; and prokaryotes: bacteria and cyanobacteria; free enzymes in clay colloids, etc.; nematodes: various invertebrates, such as earthworms and insects, etc.). By means of an *unbroken, interrelated chain,* plants take their nourishment from *specific micro-organisms which they have attracted* by biochemical messages from their root systems. This is done in such a way as to help their own development, thus creating a true *fauna-flora symbiosis,* which is as beneficial to the host as the hostee.

As it happens, root nodules play host to a whole society of micro-organisms which are *completely specific to each plant species.*

Their activity is eminently favorable to the plant, while at the same time forming the ideal development medium for invaders. These find their nutritional niche in the said nodules, whilst the plant itself can pick up the nitrogen compounds resulting from their digestive processes and decomposition.

Worth mentioning in this context are two types of complex molecular dialogues which have led to an entire array of attractions between the plant and its host: rhyzobian symbiosis and micelial symbiosis.

1) Rhyzobian Symbiosis

The first of these allows leguminous plants in particular (the vast majority of which live in symbiosis with bacteria of the Rhyzobium type) and actinorhizian plants (using their root nodules, actinorhizae, to host bacteria of the Frankia type) to feed themselves. As the plant provides its host with organic carbon-based molecules, symbiotic bacteria nourish it in return with nitrogen compounds.

2) Mycelial Symbiosis

In this second type of symbiosis, root systems use complex biochemical messages to attract filaments of mycelium which naturally enlarge their catchment area. Here too, they receive nitrogen and phosphorus, while the fungus (half-plant, half-animal) feeds off sugars produced by plant photosynthesis. Continual dialogue takes place within *mycorrhizae* ("root fungi"), supplying the host with carbohydrates and the plant with proteins.

This proves beyond any doubt that soil biodiversity is the *sine qua non* of its fertility, **creating food systems with self-sufficient closures.** Moreover, these increase the prospection area of root systems.

3) Hydric and Respiratory Symbiosis

Similarly complex symbioses regulate respiratory and water supply mechanisms of the unitary "soil-fauna-flora" chain. We should stress the **instinctive balance of the soil's three solid, liquid and gaseous phases.**

31

Water is conveyed from bacterium to bacterium and rises through the water table from the bedrock. Therefore, the flexibility and fluidity of this hydraulic system depends on the humus's biodiversity. What is more, this chain comes about in perfect co-existence with the gaseous supply feeding surface bacteria with air.

Once again, we must highlight the *spontaneous holism* of all fine tuning carried out by this double food and respiratory chain. Any human intervention, however superficial, can only be deemed as inappropriate and deeply destabilizing, on a scale ranging from futile to extremely harmful.

4) The incalculable contribution of fauna: from insects to birds and mammals...

Furthermore, all fauna is involved in plant reproduction and fruit-forming.

Numerous natural pollinators have harmoniously complementary roles, whilst predators provide an indispensable, thorough cleansing function...

32

First, let us consider the efficient relationship that develops between the ruminant and the meadow on which it grazes. We can see them cooperating and sharing straightaway – with plant and animal characteristics developing together!

Entomogamous plants depend on an infinite variety of insects to pollinate them: multiple species of butterflies and moths, various apoidea, ants, thousands of sorts of dipterans (flies, syrphoidea and bombyliidae – bee flies – etc.), the entire coleoptera (beetle) order (rose chafer, bee beetle, false oil beetle, scarab beetle, firebug, ladybird, and so on).

Through the "pollination syndrome," a vital dialogue is created with a plant, which develops an entire body of alluring characteristics for the pollinating species it wants to attract (floral structure, width and depth of the corolla, temperature, fragrance and color, etc.). At the end of the chain, we can even find strictly dual relationships in which only a single pollinator may fertilize a single plant species. Talk about the importance of *coevolution* in entomogamy!

This is based, for the most part, on the incredible diversity of apoidea (domestic and wild bees, bumble bees, wasps, mining bees, leafcutter bees and other osmia). All of these carry out fertilization of almost 80% of fruit and vegetables, in perfect synergy, and through strictly encoded sharing interplay!

Working in ideal symbiosis with its successive hosts, the pollinator gathers nectar over long distances, while the flora, for its part, provides it with an abundance of pollen and nectar, its staple food.

Consequently, fauna has an undeniable influence on a biotope's biodiversity. In the forest, many species depend on wild bee activity to perpetuate themselves through cross-pollination. Some examples include rosaceae (checker tree, hawthorn, dog-rose, sweet cherry, mountain-ash, etc.), ericaceae (blueberry, heather, etc.), lamiaceae and herbaceous plants (sages, Viper's bugloss, orchids, etc.).

An identical kind of coevolution is at work with ornithophily or bird pollination (for instance, hummingbirds, souimanga sunbirds and honeyeaters from tropical and sub-tropical regions). In this particular case, throughout their history, flowers have lengthened and narrowed their corollas and changed from yellow shades, which were adapted to insects, to red hues which can easily be seen by birds. This has been done in such a way as to adapt to the perception and beak shape of their bird pollinators.

Finally, a similar, perfect kind of mutualization occurs in chiropterophily or bat pollination. At night, flowers emit ultra-violet light which bats can see.

The most varied types of animal help disperse and transport seeds simultaneously, increasing biodiversity immeasurably. Rodents pile up acorns, beechnuts and hazelnuts; ants spread seeds, only taking interest in the oily parts; seeds with hooks or burs cling to furry mammals; berries protect seeds which can be found intact in the droppings of birds which have consumed them; and so on. This is not to forget the considerable role played by the elements, such as the wind (anemogamous plants), snow and rain.

Faced with the decline of pollinators and wildlife caused by the prolific quantities of insecticide sprayed on crops, the farming sector can only expect to see its yields dwindle inexorably to the point that an unprecedented world famine will ensue!

This is because flora is predominantly dependent for its survival on the wildlife with which it creates close symbioses!

Chapter Three
The Need for Drastic Change in the Agricultural Sector

In view of the dire need for food security, the hunter-gatherers, which we were, began looking to affect the landscape in order to ensure vital supplies. That was the dual beginning of crop growing and livestock farming which allowed us to settle down whilst providing us with an element of ease.

That is why we have witnessed the *gradual, massive destruction of Earth's primordial landscapes* throughout the course of human history. It has been a catastrophic loss because these are the only environments which can tell us about the *spontaneous balance created by the vast organism that our planet comprises.*

Furthermore, since modern times, out of greed we have substantially poisoned the air, water, soil, as well as the flora and fauna with treatments which are both ineffective – because they break the long multi-cooperative chains – and highly polluting in the long term (we do not have any antidotes!).

With rigorous observation, we should have learned lessons from these untouched landscapes, if any remain!

This is because the Earth reacts constantly in a *unitary, communicative and interrelated* fashion, and these primary equilibria were the best indicators of the inherent nature of each of its landscapes and the structures which suited them best.

Changing these fundamental landscapes is like playing God, because *we rush in without any reference points*; without any true objective other than that of the strictly money-motivated kind; with solely the venal concept of "yield" in mind.

Yet, what is yield if it is not the *health* of a thriving structure harmoniously connected to the surrounding society, which is obviously harnessing all realms?

If we keep on changing the landscape at our whim, we will destabilize the deep substratum just as we have the atmosphere and climate. We do not know what we are doing when we meddle with any of these phenomena and we are powerless to mend them.

That is why, as part of a more positive outlook, we should ask what the agriculture of tomorrow could be like?

First and foremost, it should result from a *tremendously reverent observation* of natural mechanisms. Following on from this, we, as faithful servants, will join with vital mechanisms in order to enhance them – if need be – but never spoil them.

This was magnificently achieved by primordial civilizations the world over, who were still instinctively imbued with a powerful ecological perception which modern Man has lost by completely severing ties with his roots.

Serve the earth, fauna and flora! Domesticate, of course, but in full consideration of *mineral, plant and animal intelligence* as the *natural equilibria* resulting from a *global, universal consciousness.* Compared with this, we are

39

barely any better developed than the mass of cells of which we are made. We still have not fully understood the highly relevant *"holos,"* and we cannot throw its machinery off balance for the sake of our egocentric ends without immediately suffering the consequences. Aside from the impairment we cause to our bodily envelopes themselves, the main penalty will be our irreversible removal before long from a planet we will have overly mutilated and which will utterly reject us in one fell swoop!

Chapter Four
Spontaneity and Symbiosis

Let us regain something of the spirit of those first human populations whose actions only took place within the parameters of *colossal passive equilibria* which they had repeated occasion to experience – often to their own disadvantage!

Observation will take the top priority in our activities and restrain them to the right proportions, provided that they always strictly abide by *planetary vital force*. (This reaches well beyond the Earth and encompasses the Solar System and billions of galaxies which only arise in close interdependence.)

Accordingly, it will be the sort of agriculture which takes into dual consideration *the macrocosm and the microcosm*. It will be integrated into these closely interrelated spheres. These include planetary influences – especially the Moon – and extend to the paramount conditions dictated by the micro-organisms populating the humus.

41

And, of course, this includes all of the free mineral, animal and plant interactions which take place at our scale.

Everything communicates and intermingles continually. It is precisely at the root of vital energy that a ***powerful unitary organization*** is created whose almost unlimited impact entirely surpasses any infrastructure we could pit against it.

Consequently, the more our domesticated animals live a free existence without fetters, as a gesture of love we will make to them, the less we will make them suffer (in particular, by reforming their transport and slaughter conditions), and the more our meat-based diet will be fit to ensure our good health.

Most of our diseases will disappear!

The same goes for plants. The more we rely on natural communities – mineral, animal and plant simultaneously – and retain these instinctive relationships, the less we will stress the soil. The more we respect spontaneous plant cover, contributing to the free vital impetus of all our love, the more

42

the fruit, vegetables, leguminous plants and cereals we put on our plates will be healthy and apt to keep us in good health!

Let us recall the well-known phenomenon of *dormancy*. The soil carries a profusion of seeds from various species. They only come out of their lethargic state if the composition of the substratum itself and climatic conditions, etc., allow them to do so, thus reflecting excesses and deficiencies, and attesting to the characteristics of the soil's microbial life.

Plant seeds have experienced incessant evolution for eons which has led them to enrich their substratum, in the same way that feedback from the latter allows them to germinate or not. Seeds come up *as soon as the biotope's characteristics fully meet their genetic heritage.*

It is for this reason that self-propagating plants are the best markers of a soil's nature and dynamics. At any given time, they wait for optimum conditions for them to propagate; and at the same time, they are themselves automatically regulated. In the

43

end, their demise is due to the depletion of the resources which led them to grow in the first place.

Only strict *symbiotic* agriculture, which respects natural mechanisms without forcing anything, can sustain a biotope's basic health and legitimately deserve to be called "organic."

Through it, we shall renew healthy ties with the planet which supports us.

Chapter Five
By Way of Conclusion

Every type of plant cover has an essential dynamic role for the evolution of its substratum, just as *every type of soil induces or inhibits the growth of its fauna and flora.* This naturally engenders a continuous narrative from a seamless dialogue.

We should work *with* a landscape, never *against* its innermost nature, i.e., *with vital energy and not using deadly means.*

When we act *in keeping with natural symbioses,* we *foster life.* We help augment vital energy immeasurably and this will give it back to us in abundance.

Brief Bibliography

Le Sol Vivant: Base de pédologie-biologie des sols. Jean-Michel Gobat, Michel Aragno, Willy Mathey, Collectif, René Bally (Préface). PPUR Édition, 2010.

Biologie du sol et agriculture durable. Christian Carnavalet. France Agricole, 2015.

Permaculture: Guérir la terre, nourrir les hommes. François Léger (Postface), Perrine Hervé-Gruyer, Charles Hervé-Gruyer, Philippe Desbrosses (Préface). Actes Sud Editions, 2017.

La révolution d'un seul brin de paille: Une introduction à l'agriculture sauvage. Masanobu Fukuoka, Wendell Berry (Préface), Bernadette Prieur Dutheillet de Lamothe (Traduction). Editeur Guy Tredaniel, 2005.

www.ingramcontent.com/pod-product-compliance
Lightning Source LLC
Chambersburg PA
CBHW081644220526
45468CB00009B/2545